Genetically Modified Foods Questions and Answers

Carlton Stewart

Copyright © 2017 Carlton Stewart

All rights reserved.

ISBN- 10: 1544803052
ISBN-13: 978-1544803050

TO ALL CONSUMERS ON EARTH

PREFACE

This book was written from the viewpoint of a Consumer Advocate, to satisfy an identified need for answers to basic questions on genetically modified foods, which prior to this, has not been available in this format to Consumers.

The Reader will not see or be challenged by chemical formulae or any unnecessary scientific terminologies.

The book provides concise easily read answers to relevant questions, whilst pointing to areas in which positive action from Governance is urged, and others where the reader might, if so desire, engage in deeper research.

The writer is not anti-science, but is insistent that products and services should be rigidly safety assessed and properly labeled before it is introduced into the environment for consumption of animals and human beings.

This effort is not intended to defame, slander or libel anyone but for informational purposes. We have provided some information from third parties and we are receptive to being corrected where we have inadvertently erred.

CONTENTS

1. What are Genetically Modified Foods? 01

2. Were they tested? 20

3. What are their effects? 28

4. Are there laws? 33

5. What have we done? 36

6. What should the Government do? 38

7. What can YOU do? 40

8. References i

ABOUT THE AUTHOR

Carlton Stewart is a graduate in Management from the University of the West Indies and has a Certificate in Marketing from the College of Arts Science and Technology.

He is a Consultant and Business Service Provider and has traveled extensively to markets in the Caribbean, Africa, USA Britain, Haiti, Belize and Puerto Rico.

Mr. Stewart is a Lecturer in Sales /Marketing and Consumerism. A member of the H.E.A.R.T Entrepreneurship Lead Group and the Credit Union's Credit Committee.

As a Consumer Advocate, former President of the National Consumers' League and founding executive of the Caribbean Consumer Council he has presented papers on Corporate Social Responsibility (CSR ISO 26000).

Intellectual Property Rights and Genetically Modified Foods to the Scientific Research Council and on the Terrorism Act, Credit Reporting Act, Tax Reform to Joint Select Committees of the Jamaican Parliament.

ACKNOWLEDGMENTS

References:
www.fda.gov/food/foodscienceresearch/biotechnology/submissions/ucm225043.htm#out1

www.livinghistoryfarm.org/farminginthe70s/crops_13.html

GM Foods the facts and the fiction Maria Elena Hurtado
Agriculture at a crossroads-Executive Summary of the Synthesis Report- 2009

en.m.wikipedia.org/wiki/Consumer_Bill_of_Rights
The Convention on Biological Diversity Rio de Janeiro, 1992

Reactions to the Convention on Biological Diversity
en.m.wikipedia.org/wiki/Cartagena_Protocol_on_Biosafety
https:bch.cbd.int/protocol/Supplementary/

1 CHAPTER

What are Genetically Modified Foods, and how are they produced?

These are foods that have been altered in unnatural ways.

Genes from related and non-related species are combined or transferred into living things in attempting to give them features/traits they never had before, for example fish genes are put into tomatoes to enhance resistance to cold temperature, genes are put into potatoes, corn, soya beans, cotton, canola and other crops enabling them to exude pesticides on their own, that means a pesticide comes from the plant and it's leaves which kills the pests that eats the food crop.

Redesigning Life and Nature

Crops are engineered to be resistant to herbicides and are able to survive the herbicides, in fact "drink it up"; whilst the herbicide kills the weeds surrounding the crops as is their purpose, without ostensibly harming the crop. These engineered traits or abilities produced in the crops are carried into the foods developed from these crops.

So when human beings or animals consumes these foods they are affected by how these foods are made and the chemicals which are contained in them, seen or unseen.

One way to look at foods is that they are containers of water, so whilst we may not be able to see the water in them, it is there, and its quality makes up the quality of the food. So we know that the method of production and the chemicals contained in the foods represents a major part of its quality.

Studies in the U.S.A and the Netherlands show that our babies are depositories of hundreds of chemicals which come through the umbilical blood of their mothers, during the baby's development in the placenta. These chemical can and does affect the endocrine system of the developing embryo and can cause serious defects in the baby. These defects are generational and so are passed down to our children and their children into eternity.

Genes are strings of chemicals called "nucleic acids" in the bodies of all living things. They are the units of heredity passed down to us by our parents 50 percent from each parent. Characteristics such as our appearance, abilities, even the diseases to which we are susceptible we receive through this source.

However babies are being genetically modified and at least 30 modified babies were reported to have been born in the USA in 2001. The majority of these babies developed defects and that is as much as we know about them. The Food and Drug Administration (FDA) stepped in and asserted jurisdiction over how human cells can be used. BBC News I SCI/TECH I Genetically altered babies born.

Reports in 2016 indicated that a boy child was born in the United Kingdom with three parents, three different DNA using a mitochondrial replacement technique, which permanently changes the human genome with uncertainties for health, growth, development, aging and progeny among other yet unknown functionalities. Mitochondria are like energy transformer, they provide the cells of our bodies with power/energy.

Ostensibly this will allow parents with rare genetic mutations to have healthy babies. Concomitantly the possibility of producing 'super babies' could also be what is being pursued, but as is the situation with genetically modified foods that we consume a lot of unexpected and negative consequences will develop. https://youtu.be/lgX6p3n5xtw

Genes, DNA and Genome

Deoxyribonucleic Acid (DNA) is organized into stretches of genes, some that turns on and some that turns off and still others that Scientists do not know what their purposes are.

The DNA was isolated in 1869 and is the molecule that encodes biological instructions.
The Genes coined in 1909 are called the units of heredity, inherited variations.
The Genome is the entirety of an organism's hereditary information.

So the type of organism that we are, the way our body operates, the way we look , even the sicknesses for which we have a high propensity to acquire, may have been passed on/ inherited from our ancestors.

That's why our children looks like us and behaves like us. Equally that's why cattle looks and behaves like cattle, birds like birds, dogs like dogs and plants like plants.

Organisms have thousands of genes working individually and collectively at the same time, so it becomes complicated for scientists to isolate the particular genes which embodies the traits/ characteristics that is required, and to bypass all the natural defenses of the receiving organism, to insert the genes and very importantly to get the inserted genes to interact beneficially with the body of genes into which it is inserted.

Our bodies and most organisms are gloriously created by the Almighty to resist and defend against most organisms or chemicals, virus, bacteria that are considered intruders or foreign to it. But scientists have made, and are making chemicals and products that breeches these defenses.

Genes seem to operate like a live orchestra always performing, there is intercommunication, sensitivity and responsiveness to each other and to the whole, and so the same gene may express a different behavior from one organism to another.

Epigenetics represents a relatively new field of study involving environmental effects on the behavior of our genes.

Consequently ones DNA though representing ones predisposition, is not ones destiny, as the environment factors such as the foods we eat, the exercises we do, the chemicals we are exposed to and so on, seem to have major effects on the behavior of our genes which determines how we look, the illnesses we suffer our life style and wellbeing.

Let's assume for the moment that the Dance Band "Byron Lee and the Dragon airs", a popular Jamaican band is the genome of an organism and its band members are like the genes, playing and interacting nonstop.

Now take out the gene representing the drummer whilst he is still playing, and force him into a Jamaican Rhumba Band, that is also playing continuously.

Most likely the music now coming from the Rhumba Band with the inserted drummer/ gene playing the same way he was playing whilst in his original band, the Byron Lee Band, will not be properly syncopated and could sound disgusting even sickening.

Though this analogy is simplistic it provides an indication of what can and does at times occur when genes are transferred from one organism to another, like pesticide genes in corn or soya beans. Some genes can stop functioning in the new organism, into which it is placed, or functions in unexpected ways such as promoting toxicity in the corn, soya beans or any food or drink that we feed our children ourselves or our animals. The great geneticist Dr. Mae Wan Ho provided the concept of genes behaving like quantum jazz, each one playing its own part, while at the same time interacting with each other.

Almighty God says in Genesis 1:29 Behold, I have given you every plant yielding seed that is on the surface of all the earth, and every tree which has fruit yielding seed; it shall be food for you.
Seeds were created with the inherent ability inscribed in its DNA to become trees, fruits, forest, more seeds, more trees, more fruits, and more forest.

'Genetic use restriction technology', called 'terminator technology' is used by some seed companies to make seeds sterile, thus ensuring that Farmers have to repurchase seeds for each crop that they plant. One wonders though, if our seeds stop regenerating, won't it be just a matter of time before humans and animals who are consuming the crops from these seeds stop regenerating also?

Our ability to have children, and indeed the very sustainability of the human race could potentially be at risk with the advancement of this technology.

Also technology fees and royalties which are applied to the selling price of the genetically modified seed can bankrupt Farmers who have invested in the GM technology. "In less than two decades, cotton seed has been snatched from the hands of Indian farmers by Monsanto, displacing local varieties, introducing GMO Bt cotton seeds and coercing extravagant royalties from farmers. Since Monsanto's entry into India in 1998, the price of cotton seeds has increased by almost 80000 %(from 5 rupees- 9 rupees/kg to 1600 rupees for 450 gms. 300,000 Indian farmers have committed suicide, trapped in vicious cycles of debt and crop failures,84% of these suicides are attributed directly to Monsanto's Bt cotton". Dr. Vandana Shiva

The Process

There are two basic ways in which these crops are engineered: One way is to shoot the gene into the cells of the plant using a gene gun, this is to overcome the natural defenses of the plants or animals to the foreign genes.

An antibiotic marker gene is attached, to determine which cells the genes got into, and a gene construct called a viral promoter is also attached to ensure that the genes offering the traits becomes active when inserted.

Another way is to infect the cells of the recipient organism with the gene grown in bacteria or virus until it becomes integrated. The plants successfully impacted are cloned and will contain the genes scientists' hope, which will exhibit the characteristics they are trying to induce.

Regardless of the method used in transferring the genes and the cloning there are damages to the DNA of the receiving organism which can lead to toxic products being formed and unexpected results may be produced.

Contaminating potential and its consequences.

The chemicals / viral promoters attached to the genes being transferred are promiscuous, that is they are specifically produced to promote the bypassing of the natural defenses of other organisms in the environment.

Genetically Modified Foods Questions And Answers

Consequently genetically engineered crops and trees from which genetically modified foods are obtained, contaminates organically and/or conventionally grown crops and trees. This happens by the wind blowing spores, or pollinators moving between crops thus facilitating the exchange of DNA. It is therefore not possible to plant GMO crops beside organic or conventional crops without them being contaminated. https://www.grain.org

Organically grown produce must satisfy approved standards by a body accredited by IFOAM Organics International or some other recognized certifying accreditation organization for them to be accepted as organic.

Farmers worldwide traditionally save seeds after each crop for replanting the next season and to help out a fellow farmer whose crop may have failed and needs assistance to replant. Companies like Monsanto and others who were really chemical companies but have migrated into seed production and life sciences (which is one of the ways they describe what they do) have genetically engineered seed, as we mentioned above, so they produce crops with traits, essentially two, herbicide resistance, where a farmer can flood his field with the herbicides the company produces and then plant the herbicide resistant seeds.

The seeds will grow, but nothing else will, at least in the beginning. However the experience is that after approximately two years, insects in the soil and the weeds become adjusted to the poison and are able to survive, causing even greater problems requiring stronger chemical poisons to eliminate.

Farmers can find themselves on a chemical treadmill, or like junkies requiring fix after fix, having to find stronger and stronger chemicals and at times having to removing some weeds manually. Secondly there is the pesticide producing trait, this is where the pesticides are produced and comes from the crop itself, and so once the insect eats it, they die.

One of the central agreement of the World Trade Organization (WTO) is TRIPS trade related intellectual property rights, which seeks to grant monopoly rights over life forms like seeds, animals and their offspring for generations, to commercial interests.

Under this agreement corporations can patent these GMO seeds, whilst farmers have to enter into a technical agreement with the corporation in order to purchase seeds for planting.

Consequently the farmer can be sued by any of these corporations for patent infringement if plants are found growing in a farmer's field and he does not have a technical agreement with the corporation.

This has serious implication for developing countries like Jamaica where our biodiversity (the variety and quantity of our plant species and ecosystems), should have, and can still make us a power house in the medicinal and nutraceutical industry.

Hundreds of farmers mainly in America have been sued by Monsanto but not all have lost. Percy Schmeiser a Canadian canola farmer was sued in 1998 for patent infringement after his field was found to contain Monsanto's patented GMO Canola. He decided to fight and won in March 2008 when Monsanto agreed to pay for cleaning up his field.

Monsanto GMO products have been found in areas where the planting of GMO crops are forbidden, this has happened in Brazil and led to the temporary authorization of the 2003 GMO soy six-million-ton crop by the Brazilian Government, estimated at 300,000 US Dollars

Also in Paraguay where a ban on GMO crops existed, contamination led to the eventual authorization of GMO soy in 2004.

India had a similar experience where the contamination of their cotton crop by Bt cotton in 2002 leading to their eventual approval. Contamination of organic and conventional crops by GMO crops seems to be a major strategy for the legalization of the GMO crop technology.

The Biotech Corporations have been accused of using unethical and irresponsible advertising campaigns to gain the confidence of farmers. And aggressive marketing strategies in India including educational programs in schools in Brazil to promote the GMO crop technology.

Because farmers can no longer save the seeds they buy from these Corporations they have to pay a surcharge on top of the price of the new purchase of seed which is called a technology fee and sign a technology use agreement upon each purchase which stipulate that they are prohibited from saving any GM seeds for replanting.

Health Risks

When we consume foods produce from these processes it is possible for our health to be impacted in the immediate term involving rashes, allergies possibly leading to anaphylactic shock, stomach and bowel discomfort and muscle pains could also present.

In the longer term we could see premature ageing, cancer of the organs of the body like the thyroids, liver, kidney, colon, intestine, breast, immune and infertility problems, increase in diabetes and heart diseases.

Basically the effects tend to be chronic as oppose to acute, this hides the danger, because people will not consume these foods and fall down in the streets immediately, however, over time they become critically sick, since there is no labeling, it is almost certain no one will connect the dots.

These occurrences are what is seen in the animal studies that independent research scientists enlightens us about, although humans are believed to take a longer time generally to be impacted, than animals.

From the beginning of recorded history man has used selective breeding to influence the quality of animals and plants achieved by the genes they pass down through their offspring.

An excellent example of this in Jamaica is the work of Dr. T.P. Lecky who devoted 70 of his 90 years to the development of cattle best suited for the Jamaican condition.

Genetic engineering the process that produces genetically modified foods has dramatically speeded up the process by transferring genes not only within but also cross different species and kingdoms, for example human beings into food, fish genes into crops, producing chimera from different animals and so on.

Rat insulin, human insulin and other products in the medical field were some of the first results achieved by researchers of Harvard University and Genentech using the genetic engineering technology in 1978, before the technology was used to produce crops and foods in1982.

A good analogy is to look at the DNA of species as rivers flowing through time with high banks separating and confining the species. (Richard Dawkins).

The Genetic engineering process is bursting the high banks and mixing the gene pools with the unfathomable potential for wide spread unrecoverable degradation, resulting in serious health problems in humans, animals and the environment.

With this ability to move genes around, and without labeling, we are at the mercies of scientists and those they work for in regards to what ends up in our field, feed and on our tables as food.

There are emerging gene technologies, one of which is called the "gene drives" which uses gene editing techniques to ensure that an altered gene spreads throughout an entire progeny population.

CRISPR TECHNOLOGY developed by Jennifer Doudna can change the DNA of most organisms, and have by making a small change in the DNA of black rats made them produce progeny that are white.

Genes are carriers of traits essentially, so being able to transfer genes hence traits into organisms and progeny of organisms within generations represents awesome power. We know that power corrupts and absolute power corrupts absolutely. So there needs to be transparency, fulsome debates and democratic decision making on what is done with this and other existing and emerging technologies.

Consumers' interdependence

Birds, bees, flowers, trees, animals, human beings all things that lives and breathes and uses the resources of the planet are consumers.

Carlton Stewart

Although our emphasis is mainly on human consumers, we would like to see the development of a new mindset geared to ensuring that our activities consider the interconnectedness of all existence on Earth.

Amazing, is an understatement when describing how as human beings we fit ourselves into categories and our actions and desires become conditioned by that, completely disregarding our duty to the preservation and proper functioning of the most important category consumer.

Governments, business owners, managers, scientists, lawyers, doctors, workers and various categories of consumers when making decisions do not usually consider what is good for consumers on the whole despite the fact that they themselves are consumers.

It comes down to what is good for the Government, or what is good for the business, or the particular category in which they find themselves, and here is where some of our biggest problems begins.

Currently there is a tremendous Bee die-off in the United States of America, and if this is not corrected quickly many of the foods that we take for granted could soon disappear from our supermarket shelves because the bees will not be there to pollinate the fruits and vegetables.

Another example of our interdependence is demonstrated by the dung beetle, which is found in some habitats. The dung beetle collects the dung of large animals it lays its eggs in the animal's dung rolls it up and buries it, eventually the beetles hatch from the dung, however also important is that the dung contains the seeds of plants and trees, which, because they are buried, they germinate and grow, giving us food.

Genetically Modified Foods Questions And Answers

In many areas the earth is unable to keep up in the struggle to regenerate. The International Union for Conservation of Nature (IUCN) says among other things that many species are threatened with extinction, 75% of the genetic diversity of agricultural crops has been lost.

We are not putting forward an argument against self-interest; on the contrary, we are only calling for a re-examination of self. We are nothing without each other, there is no me without you.

So whilst we do things and make decisions to satisfy our manager/leaders, constituents, or just greed, we must before signing off on any decision determine, what effect it will have on the 'collective self' other consumers and the earth to which we are inextricably linked.

If consumer, the dung beetle, became extinct then consumer, human being, will be affected eventually by having less food available to us.

As more animals and plants become extinct our biodiversity is negatively affected, but importantly, diversity is what is needed not 'Mono Culture'.

Fallacy of over population

What is this? Genesis recounts the creation story and throughout this story God made the animals, each after their kind, from experience even though you have the kind called dog, there are diversity within this category and this diversity is enhanced by the dogs having sexual relationships and producing puppies possessing genes from both parents- this is natural.

Scientists supported by large research budgets produced by the rich, who's motivation really is the huge profit potentials of making life that they can patent (IPR), intellectual property rights and of course the opportunity to feel like Gods should not be discounted, have for twenty years or more been interfering with the genetic structure of consumer animals, insect and plants to produce something that the world will be willing to pay for, and to make as much of them as possible, thereby enriching themselves.

Another view is that the motivation is coming from World Governments having covertly decided that they need to depopulate the Earth; consequently they have allowed and encouraged the development of experimental endocrine disrupting products to reduce and prevent population growth, weaken our immune system and reduce our life span.

Our Endocrine System consists of: Pineal body, Pituitary, Hypothalamus, Parathyroid, Thyroid Gland, Adrenal glands, Thymus gland, Pancreas, Ovaries, Testicles.

Now, foods and chemicals which disrupt this system disrupt our whole being. The endocrine system is deep; it is the entire workings of our body. Many people find it hard to understand how the food and chemicals we consume can cause us to behave in the negative ways some are behaving, and how these GMO foods and chemicals are generating debilitating illnesses and death.

Again, this is contrary to Almighty God instructions "Be fruitful and multiply" in Genesis 1:28
As at the world census figures of December 10, 2008 there were 6,867,020,300 people living on earth, the earth's solid surface is 36,800,000,000 total acres of land to be divided up. Each and every person on earth could receive 5.36 acres of land easily.

"The world is not short of food either, current production if equally shared, could provide everyone, each day with more than two pounds of grains beans and nuts; about a pound of fruits and vegetables; nearly another pound of meat, milk and eggs" the organization Christian Aid says and reminds us that "hunger is political.

People are shocked to discover that during the worst period of the 1984 famine in Ethiopia, some of the best farming lands were being used to grow animal feed for export to Britain and the rest of Europe. In 1995, India exported five million tons of rice and $625 million worth of wheat and flour, at the same time one in five Indians went hungry"

Unregulated GMOs

The first commercial genetic engineering of a GM plants was done in 1982 when Flavr Savr tomatoes were developed containing genes from the flounder fish. Potatoes were engineered with genes from silk moth, viruses and bacteria. By 2006 80 million hectares of GMO crops were planted around the world primarily in the United States, Argentina and Canada. According to industry funded group ISAAA (International Service for the Acquisition of Agri-Biotech Applications) in 2013 there were 175 million hectares in GMO crops, America, Latin America, Asian and African farmers are the main areas where they are planted. America has the most GMO crops planted where three crops corn, cotton and soya bean makes up the majority of the 68.4 million hectares they have planted.

Based on (FAO statistics for 2009) GMO crops occupy 11.45% of cultivable land globally, compared with 88.55% for non GMO. GMO crops are confined to 27 countries while non GMO crops are planted in 196 countries.

Since the Flavr Savr introduction, 'Pandora's Box' has been opened, human genes have been put in food, and anything goes

Bovine Somatotropin BST a genetically engineered hormone is injected into cows in order that they will produce greater quantities of milk, 10 to 15% greater.

But in practice, many cows become very sick and produced puss, blood and feces contaminated milk. This hormone is associated with prostate, breast and gut cancer in humans, but without labeling of the milk, liability cannot be established.

There are lots of new products being formed that nobody knows for certain what their effects are likely to be.

Many people were inflicted by an excruciatingly painful disorder linked to a bacterium engineered to produce the food supplement L-tryptophan, an essential amino-acid present in normal diets.

However, one producer artificially inserted genes into a bacterial species to produce the substance and in late 1989; ten thousand people fell ill with an unusual illness eosinophilia-myalgia syndrome (EMS), caused by Showa Denko tryptophan, within months dozens of people died and thousands maimed and thousands suffered from excruciating pain even today.

The information is that because of the severity of the attack, the problem was traced to an unexpected outcome of that genetically modified product.

Genetically Modified Foods emanates from the process of genetic engineering whose proponents exuberantly announced from early that it can produce limitless benefits to the field of Agriculture.

They argue that they are going to be needed to feed the world and that they, have the potential to provide crops with all kinds of traits for example, tolerance of drought, cold, salinity and flooding, resistance to insects and pests, extra nutritional value and more.

However since 1998 despite many Biotech companies being established and billions of dollars spent, they have failed to realize these exuberant claims, with only the establishment of two main traits, the tolerance of glyphosate based herbicides (one of the most carcinogenic substances known to man, that seem to be everywhere including our bodies), and the ability to produce a Bt toxin that can kill corn and cotton pests, whilst conventional breeding techniques have continued to outperform them.

Genetically Modified Crops/Foods are patentable and this represents an incentive to the Biotech firms, and based on the results of research conducted by Professor Seralini of France, Dr. Ermacova of Russia, Dr. Arpad Pusztai and others, also to the depopulation/ eugenicist groups.

Development of Biotechnology Firms

In the latter part of the twentieth century many Biotech startups were established, no doubt driven by the promise of huge potential profits.

However over the years and into the twenty first century Monsanto, Syngenta, Bayer, DuPont, Dow and BASF had gobbled up most of them. Further rationalization is imminent with DuPont and Dow announcing the possibility of a $ 130 billion dollar merger and Monsanto and Bayer confirming their merger at a cost of 66 Billion US Dollars in 2016 one of the largest of its kind ever.

Syngenta has been sold for $43 billion dollars to China

National Chemical Corp. Essentially the Biotech Companies have been undergoing major rationalization down to the Big four Syngenta, BASF, Monsanto/Bayer and Dow/DuPont.

Monsanto was started in America in 1901; it is the maker of Roundup the largest selling herbicide, with its main constituent glyphosate. There is ambivalence as to its carcinogenic and autism causing effects. It was recently found in the urine of 93% of Americans that were tested, over 280 million pounds were used in America in 2010; along with Dow Chemicals, Monsanto was the largest producer of Agent Orange a defoliant that killed and maimed many Vietnamese, and US veterans. Even today it is believed that Vietnamese children are still being born deformed because of these chemicals.

Monsanto was reported to be responsible for the carcinogenic PCBs industrial chemicals banned in 1979, and DDT pesticide which was banned in 1972. This powerful Corporation is believed to have helped in producing the atomic bomb that featured in the 2nd World War. In 1985 Monsanto bought GD Searle who had the patent for Aspartame artificial sweetener which was a known neuro toxin, made from the feces of genetically engineered bacteria, now over 6000 products contains this sweetener. Monsanto also created rbgh growth hormone to increase the production of milk by cows that we mentioned earlier.

"A firm called Epicyte developed an Epicyte gene which when it is eaten by either males or females and they have sexual intercourse, antibodies from either of the partners attacks the sperms rendering them unproductive and infertile. The firm was bought by Monsanto and DuPont, and the gene was commercialized". Dr. Rima Laibow

MSNBC uploaded a video clip on Dec 7, 2010 from the show

"Scarborough Country "explains how the Bayer Pharmaceutical Company has knowingly been distributing an HIV contaminated drugs to hemophiliacs for over ten years, and after being exposed for doing so they shipped the remaining stock to be used overseas. Bayer also produces chemicals which are harming Bees and Worms.

These are the main corporations that are moving to control the agricultural system of the world. And, it is suggested that the best way to control a people is to control their food.
http://bestmeal.info/monsanto/company-history.shtml

.

2 CHAPTER

Were these "novel foods" tested and safety accessed, and by whom?

One would believe that before the introduction of a category of food that is new and unusual involving the transference of DNA from one specie to another. Before this is presented on the market for people to buy, consume and give to their families, that it would at least be safety assessed. To include animal and human studies by both the company producing the food product and most importantly the institutions which approves the food's introduction into their markets?

What has happened however is that the institutions like the Food Drug Administration (FDA) of America, which gives the approval allowing the "novel foods" to be placed on the market, they themselves do not conduct tests, but depend on the producers of the product(s) to do tests and submit the results to them. Nation states like us in the Caribbean do not conduct our own safety assessment of these novel foods; however that has to change given the multitude of these novel foods impacting our markets and the need to protect the health and well-being of our people.

Consumer protection

One cannot depend on a fish vendor to tell one when his or her fish is stinking! Some will, but most won't.

However the main concern at this point in this discourse, is the levels of risk assessment that takes place before the market is impacted, especially when you consider that the bodies especially in the USA that are responsible for risk assessment have been hijacked by the large business interests, who are likely to be more interested in safeguarding their massive investments, rather than the health and safety of consumers and or covertly promoting the depopulation agenda of the eugenicists among them.

Concomitant with trade liberalization worldwide is a transfer by some Governments to Industries of the standards setting and safety assessment functions for goods and services.

Terminologies such as "substantially equivalent" and (GRAS) generally recognized as safe" became a part of the safety assessment language.

Producers can use these phrases to disguise significant variations in product quality and further erode the already weakened assessment and regulatory functions. Strengthened by a philosophical belief that the absence of evidence is evidence of absence, consumer welfare and environmental safety is easily compromised.

Industry representatives because of their financial power are positioned, as head of regulatory bodies including international organizations whose remit is to ensure consumers protection and safety, but are employees of, or beholden to industry.

Consumer welfare and environmental safety suffer tremendous setbacks due to improperly structured boards with conflicting interest at the leadership levels.

Despite what seems like good intentions of Industry, the setting of standards and safety assessment ought not to be left up to them. Properly structured regulatory bodies must ensure that standards developed are adhered to, and where substandard products are made, the procedure exist for trace ability, recall, redress and disposal.

In the case of the Flavr Savr Tomatoes, one of the earliest example of a food product derived from using a new plant variety, developed using recombinant DNA techniques, Calgene its producer conducted a number of studies using rats.

In two of three studies there was evidence of stomach erosion in some of the rats. The FDA concluded among other things that the tomato was safe as other tomatoes, no major difference to require special labeling based on the information provided by Calgene.

It was believed however to be too costly to market the tomato using that technology, at least that is the reason put forward and in 1996 Monsanto bought the rights and patents to the Flavr Savr. The Flavr Savr experiment was taken off the market and has not returned in that form, as far as we know.

I saw some tomatoes first appearing on the Jamaican market in 1999, they looked good unblemished and different, however they were not labeled so I cannot say what they were, but after a short time they disappeared.

Major Independent Scientific Studies on GMOs

In 1995 Dr. Arpad Pusztai and his team from the Rowett Research Institute Aberdeen UK, was selected to conduct a study to assess on behalf of the Scottish Agriculture Environment and Fisheries Department the safety of genetically engineered Desiree Red Potatoes.

Genetically Modified Foods Questions And Answers

Dr. Pusztai found that the rats fed the potatoes had stunted growth and a repressed immune system and this was a result of the transformation system in the production of the potatoes, he also found that the testes of the rats changed from the color pink to blue.

He went public with the information and said that he certainly would not eat them, and that it was unfair to use our fellow citizens as guinea pigs, he was fired and attempts made to discredit him.

In 2005 Russian Academy Scientist Dr. Ermakova did a 19 day study on the reproductive ability of rats fed on genetically modified soy and non-genetically modified soy. The rats were fed 2 weeks before mating, during pregnancy and lactation with the diets.

Litters produced by pregnant and suckling female rats fed diets containing genetically modified soy had significantly higher number of dead pups 25 as against 3 in the non-genetically modified fed group. And those pups that live were significantly smaller in size, less than 50% of the non-genetically modified fed group.

In Jamaica many of our children are fed genetically modified soy directly in food and drinks. However in Europe and most of the remaining areas of the world genetically modified soy is fed to animals we have to be very very concern.

The "substantially equivalent" argument promoted by the people who produce these products and their supporter claim that as a result they do not require any special testing or labeling.

However, when they visit the patent office, they say that

these foods are new creations, which never existed before and so are patent worthy.

So the FDA does not conduct any test, and the producers of genetically modified foods are very "creative" in any test that they do.

Dr Ruby Carman an Australian Scientist after receiving complaints from pig farmers of high rate of intestinal problems in their pigs, the thinning of their intestinal wall, the bleeding out to death of some pigs, the reduced ability of the pigs to become pregnant and problems of reproductive health, all happening with pigs who were eating the genetically modified corn and soy pig feeds.

She decided along with colleagues to do a study to determine what the causes were, because the tests being conducted by the industry seemed inadequate.

So they selected 168 pigs and divided them into two groups, 84 per group, and in a blind unbiased study, one group was fed the GMO based corn and soy feed, while the other group was fed non gmo feed from birth to the time they were killed, which was five months.

Both groups of pigs stomach and uterus were examine and it was found that the gmo fed group had 2.6 times higher swollen and inflamed stomach and their uterus- were 25% heavier than those pigs who were fed non GMO rations.

There has been some independent studies done on genetically modified foods since their introduction to the market over twenty years ago, however the French Professor Gilles-Eric Seralini's chronic toxicity study on the Glyphosate/Roundup based Herbicide and a commercialized Genetically Modified Corn Monsanto's NK 603 is possibly the most rigorously peer reviewed and comprehensive study ever undertaken on this subject, at this time of writing.

This study lasted for two years and examined the effects of (1) Genetically Modified Corn (NK 603) (2) Genetically Modified Corn plus Roundup residue (3) Roundup, on rats.

It involved two hundred rats and examined them and control groups thoroughly throughout their life span, no effects were evident until after a three months period. Now the three months period is very important, because the main Industry proponents of these GMO products were doing and reporting on studies lasting only three months.

However, after three months enormous 17 mm by 17 mm tumors began appearing in the rats 94% in the mammary glands which equates to breast cancer, and the remainder equating to renal tumors for males.

The Roundup herbicide proved to have a toxic effect on the entire organisms of the rats tested.

However a major determination of the study was that the genetic modification of a food or cereal can have differing intensity of negative effects, so it can be bad or it can be dangerous.

The process of genetically engineering the plant is causing deleterious effects in its genome and resulting in the negative expression of the food.

Six times more deaths of rats occurred in those fed the genetically modified corn as against those in the control group.

Professor Dr. Tyrone Hayes a Biologist at Berkeley University in the U.S.A was asked by Syngenta in 2003 to examine the effects of Atrazine it's most popular herbicide on frogs.

In his study he exposed his frogs to a small and diluted amount of the herbicide 0.1 part per billion, about one thousand of a grain of salt in two liters, the average Farmer planting sugar or corn could be exposed to two hundred and ninety million times more.

The Environmental Protection Agency (EPA) recommends three parts per billion in drinking water, which is thirty times higher than is effective in feminizing frogs. The frogs where changed from males to females and hermaphrodites and their immune systems were severely compromised.

Atrazine, Dicamba, Glufosinate, Roundup, 2-4D are herbicides which goes concomitantly with genetically modified foods, some crops are engineered to be resistant to one particular herbicide and sometimes they are engineered to be resistant to more than one herbicide at a time.

The European Union announced a ban of the herbicide Atrazine in 2003; however it was substituted by terbuthylazine which is reported to cause testicular cancer and similar problems as Atrazine.

Dr. Tyrone Hayes along with 21 other Scientists published a paper showing that the effects of Atrazine are consistent across amphibians, fish, reptiles, birds, laboratory mammals, laboratory rodents and with human epidemiological data.

What is clear is that we are exposed to many of these

chemicals through our foods and drinking water and an increasing percentage of our people seem to be confused about their gender and have developed cancers and autoimmune illnesses.

CHAPTER 3

What effects are these "novel foods" having on health, the environment, Sustainable development and the need to eliminate hunger and poverty?

Genetically Modified Foods are inherently unsafe, inferior in relation to nutritional density, unsustainable, using up more of our resources to produce less food, is inferior in increasing yields, and diverting scarce resources from proven ecological farming methods into genetically modified organism methods.

They are creating chronic illnesses in people, animals and the environment, and their accompanying chemicals are just as destructive.

Many experiments can be stopped, or returned to the previous state if unexpected results occur. Not so with genetic experiments, once released into the environment they cannot be withdrawn, for example, fishes that are genetically modified goes out into the oceans mates with other fishes and genetically changes the other fishes, the mutation continues through eternity, animals which are fed with genetically modified feeds becomes genetically changed.

We eat them and we become changed, gmos planted on our fields releases pollen which travels over 1000 miles and contaminates, mutates, pollutes and infect people who live in the communities, our water systems, our organic and conventional crops, our animals and fields, all are changed irreversibly.

Our children are given vaccinations and baby formula, some of which are genetically modified and they are also changed, some become autistic. Irrefutable evidence showed that black or boys of African extract are three times more susceptible to getting autism from the MMR vaccines than do white boys, and that these genetically modified vaccines do give some children autism.

Used as Guinea Pigs

The injuries and damages which have occurred and dangers which continue to exist are obscured in Jamaica by our lack of environment, animal and human surveillance. The 'absence of evidence is not evidence of absence' but of our negligence in observing and establishing potential causal linkages.

The experiments that are genetically modified organisms were released into the environment some twenty years ago, and into our food supply without any human, and inadequate animal tests.

Consequently we have and are being used as guinea pigs to test these products. Tests conducted by independent scientists, and surveillance reports coming from countries like the United States Of America, Russia, France, Britain, Germany shows strong correlations between the diseases developed by rats and hamsters who have been fed genetically modified feeds and those expressed by human consumers who are inadvertently eating these foods.

It is no surprise then, that these countries, (excluding America who is the main proponent of these experimental foods), are among the more than 64 countries, (and increasing), who have banned and or instituted mandatory labeling of gmos.

Problems such as cancers, tumors, gastrointestinal illnesses, allergies, anaphylaxis, diabetes particularly in children, dementia, Alzheimer in ages between 40 and 50 years old, still births, infertility, autism spectrum disorders, mental illnesses, endocrine disruption and others are showing phenomenal increases since 1992 in countries that are looking.

One in thirty six boys are being born autistic, in less than twenty years it will be every other boy. We believe that a similar occurrence is happening here in Jamaica, but without the records we will never know to what extent.

Countries that grow genetically modified soy, corn and cotton (which are among the main crops being genetically modified at this time) are reporting serious destructive effects on their environment and people, particularly in relation to crop failures in the Philippines, and massive suicides in India.

The spraying of Roundup herbicide (a product, which is used extensively in Jamaica on conventional crops, and is the main herbicide used on Monsanto's genetically modified crops), its main ingredient being Glyphosate, a chelator and endocrine disrupting chemical which causes deformities in Argentinean children who live near the areas where the spraying is done.

Roundup herbicide is considered more dangerous than its main ingredient Glyphosate because it contains other chemicals which promote the absorption of the chemicals into the cells of plants, animals and humans.

The chelating feature of the herbicide which ensures the eventual demise of weeds is based on its ability to block the nutritional pathways of the weeds ensuring that they become micro nutrient deficient, it also kills natural useful predators within the soil and promotes the development of insects that are resistant to the herbicide, hence more and stronger chemicals are required, like 2-4-D a major constituent of agent orange a defoliant used in the Viet Nam war.

This defoliant contaminated water supplies, the environment and is responsible for monstrous deformities in children being born even today more than forty five years after. Recall that the process of genetic engineering incorporates the genes of the herbicide/ Roundup, or 2-4-D, and or a pesticide (Bt.) Bacillus Thuringienesis, or whichever, consequently every cell in the crop and the eventual food that is produced emits the herbicide and or pesticide.

Scientist have proven, by using colostomy tests, incidentally the only reported human test done with GMOs, that these foods do not just pass out of our guts but some remain, affecting our (DNA) Deoxyribonucleic Acid and the bacteria in our gut, there they continue the blocking of the micro nutrients and amino acids we need and the destruction of useful bacteria, whilst producing horrible toxins. see more at https://www.ncbi.nlm.nih.gov.

Nutrient deficiency in GMOs

We believe that most of our diseases are a result of our bodies compensating for the insufficiency of micro nutrients needed for our survival, allowing depletion in one organ for the benefit of the whole body.

Tryptophan a precursor to Serotonin which is essential to the proper functioning of our brain and body is depleted both in the foods we eat and within our gut, low Serotonin is associated with violent behavior, anxiety, and depression.

This has been proven in situations where GMO have been removed from children's diet, and replaced with organic foods, and within seven days their behavior changed dramatically, also cattle in South Africa were found to be violent, including eating each other tail, their ration was changed from GMO grains to organic grains and their behavior also changed dramatically, returning, when they were switched back to the GMO grains.

The blocking and depletion of micro nutrients by these chemical and expressed in the foods we are consuming is also believed to be playing a central role in the explosion of children obesity.

Our bodies are seeking to obtain the necessary nutrients required for its sustenance from the nutrient deficient foods we are providing it, not being able to receive enough, we keep eating more of the same nutrient deficient food in search of it, and hence we become obese.

The International Assessment of Agricultural Knowledge, Science, and Technology for Development (IAASTD) in a three year study in 2008 and represented by four hundred Scientists around the World, recommended that the focus be on building resilience and health in communities through sustainable agricultural techniques which it groups under agro ecological methods.

Dr. Rima Laibow a Psychiatrist and director of Natural Solutions Foundation, quite appropriately changed their names from Food to "Phudes" and talks about the bio-tech firm called Epicyte which developed an Epicyte gene which when it is eaten by either males or females and they have sexual intercourse, antibodies from either of the partners attacks the sperms rendering them unproductive and infertile.

Scientist are attempting to introducing gene silencing, population sterilization and pharmaceutical functionality within foods, without fully understanding how genes operate causing unpredictable consequences which an unsuspecting and unprepared consumer has to deal with.

4 CHAPTER

What are the International laws and protocols to treat with these foods?

Consumer Bill Of Rights
President John F Kennedy on the 15th March 1962 in a speech to the United States Congress extolled four basic Consumer Rights:

The Right to Safety
To be protected against the marketing of goods which are hazardous to health and life

The Right to be Informed
To be protected against fraudulent, deceitful or grossly misleading information, advertising, labeling or other practices and to be given the facts he or she needs to make an informed choice.

The Right to Choose
To be assured wherever possible, access to a variety of products and services at competitive prices; and in those industries in which competition is not workable and Government regulation is substituted, an assurance of satisfactory quality and service at fair prices.

The Right to be Heard
To be assured that Consumer Interests will receive full and sympathetic consideration in the formulation of Government Policy and fair and expeditious treatment in its administrative tribunals.

The United Nations through the United Nations Guidelines for Consumer Protection expanded these four rights into eight; The right to satisfaction of basic needs, The right to redress, The right to consumer education, The right to an healthy environment. These Rights are the foundation of the Consumer Movement, and are being abrogated by the actions of some Governments and some Corporations.

The Convention On Biological Diversity(CBD) Rio de Janeiro 1992 : December 1993 entry in force.

Conscious of the need to organize, protect, conserve and the establishment of procedures in regards to ownership of the Biological Diversity of the World, 193 nations excluding the United States of America signed this agreement in 1993.

Article 1. Objectives
The objectives of this Convention, to be pursued in accordance with its relevant provisions, are the conservation of biological diversity, the sustainable use of its components and the fair and equitable sharing of the benefits arising out of the utilization of genetic resources, including by appropriate access to genetic resources and by appropriate transfer of relevant technologies, taking into account all rights over those resources and to technologies, and by appropriate funding.

Following this came the Cartagena Protocol on Biosafety:

The Cartagena Protocol On Bio safety January 2000 adoption, September 2003 entry in force

Managing and regulating the potential adverse impact of genetically modified organisms on the environment.

To contribute to ensuring an adequate level of protection, and the conservation and sustainable use of biological diversity, taking also into account risks to human health. The Protocol sets an enabling environment for regulating Trans Boundary movement by the establishment of a set of procedures that importing and exporting countries can follow among them being the advance informed agreement procedure.

It also underscores the applicability of the Precautionary Principle, which allows countries to take conservative, risk prevention measures in the absence of detailed scientific data on the impact that a GMO may have on human health and biodiversity.

And a clearing house mechanism to promote and facilitate technical and scientific cooperation between members
A major short coming of the protocol is that it does not cover products derived from GMOs, such as processed foods, which has ingredients that come from GMOs.

Nagoya-Kuala Lumpur Supplementary Protocol on Liability and Redress adoption October 2010

The aim here is to contribute to the conservation and sustainable use of biodiversity, taking into account risks to human health by providing rules and procedures on liability and redress.

This relates to damages caused by GMOs which enters into a country knowingly or unknowingly.
Parties must provide appropriate response measures, reasonable actions to prevent or avoid damage to biodiversity and human health.

A Competent Authority must be setup by the country and domestic laws are so organized to ensure that where liability is determined, then compensation and redress is required of the operators.

In the case that operators for whatever reason are not able to provide redress, then the Competent Authority must provide the redress and claim from the operators.

It is going to be difficult if not impossible to identify liability where there is no traceability, and without labeling, there will be no traceability.

I also draw attention to aspects of the Jamaican Constitution in particular the CHARTER OF FUNDAMENTAL RIGHTS AND FREEDOM

Chapter 111: Charter of Fundamental Rights and Freedom Provides :(1) the right to enjoy a healthy and productive environment free from the threat of injury or damage from environmental abuse and degradation of the ecological heritage;

In my view the nonexistence of labels on genetically modified organisms in food, feed and field has abrogated our Constitutional rights and stymied all the United Nation Protocols mentioned above.

5 CHAPTER

How has Jamaica regulated its population's exposure to genetically modified foods, relative to actions taken by other Nations?

Government Administrations in Jamaica have always supported the Consumer Movement, always, and there has never been any pressure to conform to any particular view.

The Non- Government Organization the National Consumers' League established in 1966 and the Mother of Consumer Organizations in our region have enjoyed support from all Administrations over the years.

Genetically Modified Foods Questions And Answers

The Consumer Affairs Commission an agency of the Government works to the benefit of consumers in Jamaica.

Jamaica became party to the Convention on Biological Diversity on 6 January 1995.
The Country signed the Cartagena Protocol on Biosafety on the 4th June 2001 and has ratified the Protocol on the 25th September 2012.

We still have not managed to sign the Nagoya Kuala Lumpur Supplementary Protocol on Liability and Redress, and we still do not have a Biosafety Policy at the time of writing.

In comparison, sixty four countries and growing, have labeled genetically modified food and twenty have banned them.

Unfortunately the other countries in the Caribbean region are equally tardy in establishing biosafety policies as is Jamaica.

CHAPTER 6

What must the Government do to ensure that the rights and well-being of the Nation are protected in relation to Genetically Modified Organisms in Food, Feed, Field and the Environment?

We urge our Governments of the region to do the following:

(A) Develop and implement National Biosafety Policies which are collaborative and includes;
1) Mandatory Labeling of Genetically Modified Organisms in Food, Field and Feed.
2) A Suigeneris System to institute an appropriate form of protection of our plant varieties, traditional knowledge.
3) Precautionary Approach, in order to protect human beings and the environment from serious or irreversible damages we can refuse from doing, implementing or accepting products, services and processes.
4) Advance Informed Agreement, should require that before a Genetically Modified Organism is imported into our country and region, there needs to be (a) prior notification on the proposed import. (b) Full information about the GMO and its intended uses. (c) An opportunity to risk- assesses the GMO and decides whether to allow the import.

5) Consumer Education, decision making (a) There must be facilitation and inclusion of Consumers in decision making on GMOs and biological diversity.

6) Clearing House Mechanism (a) Mechanisms which provides effective information services to facilitate the implementation of the national biodiversity strategy and action plan.

7) Competent Authority (a) the establishment of a person or

organization that has the legally delegated or invested authority, capacity, or power to perform designated functions.

8) Testing and Risk Assessment (a) Certified Laboratories and testing facilities, processes, protocols and human capacity needs to be establish, for the evaluation of imported and domestic products, services and processes.

9) Liability and Redress Procedures (a) Mechanisms must be instituted to determine and assess damages to humans and the environment and the provision of compensation in keeping with the Nagoya Kuala Lumpur Supplementary Protocol on liability and redress.

10) Primacy of Public Health (a) The health and well-being of human beings, animals and the environment are the bench marks on which all measures to be implemented are evaluated and determined.

(B) In 1999, 35 million small family plots produced 90% of Russia's potatoes, 77% of vegetables, 87% of fruits, 59% of meat, 49% of milk and Russia only has 110 days of growing season per year.
Major agricultural land reforms are required to put more people on the land and to develop an organic agro ecological base for our agriculture.

7 CHAPTER

What must citizens do to get involved in the genetically modified foods situation?

Learn for yourself what genetic engineering is and the foods being produced with that system, and where possible avoid them, you and your household, essentially transform yourself into an informed/conscious consumer unit, that is the first thing.

Get involve with a community or national/regional consumer / environmental protection group and become active. Think of creative ways to inform your friends and neighbors about genetically modified foods and other relevant matters.

Write to your Member of Parliament and Government informing him/ them that you want these foods labeled enabling proper identification of them, which is our right.

Probably the most revolutionary and therapeutic activity one can engage in concerning one's consumption is to plant one's own food or buy from a local organic farmer.

Consumer consciousness

We can no longer afford to shop unconsciously, our lives and that of our families depend on us being informed and using our spending power to direct the path of our Nation and Region.

The producer has a store of goods that he has either produced himself, purchased locally or imported and he displays them, the goods, and offers them for us to buy.

When we buy them, he becomes able to continue the

process of producing more goods for us to buy. On a regular basis he will examine his stock of goods to see those he needs to replenish and generally what is the situation with his inventory/stock.

This is where the power to direct the path of our Nation and region lies in our hands as Consumers.

If we decide to stop buying those goods that he is producing, then he will have to stop producing them, because his inventory/stock will show that they have not moved, are not selling and soon, he will have to change what he is offering for sale, or otherwise go out of business.

So as Consumers our collective decisions determine what is presented in our markets for sale. That is the fact, and we can go a far way in protecting ourselves and family by becoming informed and making conscious decisions with our welfare and that of our families paramount. Almighty God Bless You and Your Family.

www.ingramcontent.com/pod-product-compliance
Lightning Source LLC
Chambersburg PA
CBHW061449180526
45170CB00004B/1622